Construction Zone

Diggers

by Rebecca Pettiford

Bullfrog Books

Ideas for Parents and Teachers

Bullfrog Books let children practice reading informational text at the earliest reading levels. Repetition, familiar words, and photo labels support early readers.

Before Reading

- Discuss the cover photo. What does it tell them?
- Look at the picture glossary together. Read and discuss the words.

Read the Book

- "Walk" through the book and look at the photos. Let the child ask questions. Point out the photo labels.
- Read the book to the child, or have him or her read independently.

After Reading

- Prompt the child to think more. Ask: Diggers are big machines. They dig and lift. Do you know other big machines that do this?

Bullfrog Books are published by Jump!
5357 Penn Avenue South
Minneapolis, MN 55419
www.jumplibrary.com

Copyright © 2023 Jump! International copyright reserved in all countries. No part of this book may be reproduced in any form without written permission from the publisher.

Library of Congress Cataloging-in-Publication Data

Names: Pettiford, Rebecca, author.
Title: Diggers / by Rebecca Pettiford.
Description: Minneapolis, MN: Jump!, Inc., [2023]
Series: Construction zone | Includes index.
Audience: Ages 5–8.
Identifiers: LCCN 2021053978 (print)
LCCN 2021053979 (ebook)
ISBN 9781636908557 (hardcover)
ISBN 9781636908564 (paperback)
ISBN 9781636908571 (ebook)
Subjects: LCSH: Excavating machinery—Juvenile literature.
Classification: LCC TA735 .P465 2023 (print)
LCC TA735 (ebook) | DDC 621.8/65—dc23/eng/20211208
LC record available at https://lccn.loc.gov/2021053978
LC ebook record available at https://lccn.loc.gov/2021053979

Editor: Jenna Gleisner
Designer: Michelle Sonnek
Content Consultant: Ryan Bauer

Photo Credits: uatp2/iStock, cover; Petair/Shutterstock, 1; Vereshchagin Dmitry/Shutterstock, 3; Sablin/iStock, 4; GIRODJL/Shutterstock, 5; smereka/Shutterstock, 6–7, 23br; Henrik A Jonsson/Shutterstock, 8–9; joseh51camera/iStock, 10; michaeljung/Shutterstock, 11, 23tl, 23bl; Dmitry Kalinovsky/Shutterstock, 12–13; OlegDoroshin/Shutterstock, 14; AYDO8/Shutterstock, 15; Aleksandr Rybalko/Shutterstock, 16–17; Maksim Safaniuk/Shutterstock, 18–19; Maksim Safaniuk/iStock, 20–21, 23tr; Bjorn Heller/Shutterstock, 22; parrus/iStock, 24.

Printed in the United States of America at Corporate Graphics in North Mankato, Minnesota.

Table of Contents

A Big Dig	4
Parts of a Digger	22
Picture Glossary	23
Index	24
To Learn More	24

A Big Dig

A digger digs.
It is a big machine.

A digger lifts.

It moves things, too.
It moves on tracks.

It drives over rocks and mud.

Joe will clear this land.

He gets in the cab.
He uses the controls.

controls

The boom goes out.

The arm goes down.

The bucket is big.

bucket

It digs.

It digs up a lot of dirt!

It lifts the dirt.

It dumps it in a truck.

The cab spins.

The digger digs on all sides.

Diggers do big jobs!

Parts of a Digger

What are the parts of a digger? Take a look!

Picture Glossary

cab
The area in a big machine where the driver sits.

clear
To remove things that are covering or blocking a place.

controls
The levers, switches, and other devices that make a machine work.

tracks
The steel belts on the bottom of a digger that move the machine.

Index

arm 13
boom 13
bucket 14
cab 11, 19
clear 10
controls 11
digs 4, 15, 19
drives 9
dumps 16
lifts 5, 16
moves 6
tracks 6

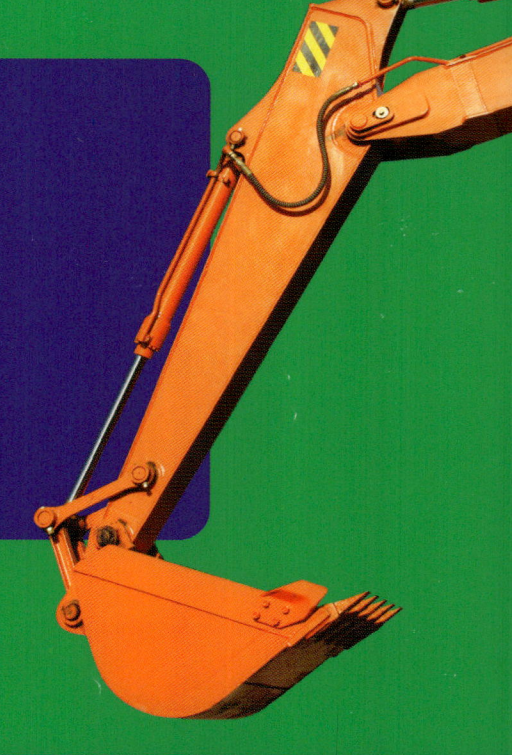

To Learn More

Finding more information is as easy as 1, 2, 3.

❶ Go to www.factsurfer.com
❷ Enter "diggers" into the search box.
❸ Choose your book to see a list of websites.